I0394249

Table of Contents

Introduction

With the trend toward a highly mobile workforce, the use of handheld devices such as Personal Digital Assistants (PDAs) is growing at an ever-increasing rate. These devices are relatively inexpensive productivity tools that are quickly becoming a necessity in government and industry. Most handheld devices can be configured to send and receive electronic mail and browse the Internet using wireless communications. While such devices have their limitations, they are nonetheless extremely useful in managing appointments and contact information, reviewing documents and spreadsheets, corresponding via electronic mail and instant messaging, delivering presentations, and accessing remote corporate data.

Manufacturers produce handheld devices using a broad range of hardware and software. Unlike desktops and notebook computers, handheld devices typically support a set of interfaces that are oriented toward user mobility. Handheld devices are characterized by their small physical size, limited storage and processing power, and battery-powered operation. Most Personal Digital Assistant (PDA) devices provide adequate memory (at least 32MB Flash ROM and 64MB RAM) and processing speed (200Mhz or higher) for basic organizational use. Such devices come equipped with a LCD touch screen (one-quarter VGA or higher) and a microphone/ soundcard/ speaker, but usually lack a QWERTY keypad. One or more wireless interfaces, such as infrared or radio (e.g., Bluetooth and WiFi) are also built-in for communication over limited distances to other devices and network access points; so too are wired interfaces (e.g., serial and USB) for synchronizing data with a desktop computer. Many high-end PDA devices also support Secure Digital (SD) and Compact Flash (CF) card slots for feature expansion. Over their course of use, such handheld devices can accumulate significant amounts of sensitive corporate information (e.g., medical or law enforcement data) and be configured for access to corporate resources via wireless and wired communications.

One of the most serious security threats to any computing device is unauthorized use. User authentication is the first line of defense against this threat. Unfortunately, management oversight of user authentication is a persistent problem, particularly with handheld devices, which tend to be at the fringes of an organization's influence. Other security issues related to authentication that loom over their use include the following items:

- Because of their small size, handheld devices are easily lost or stolen.
- User authentication may be disabled, a common default mode, divulging the contents of the device to anyone who possesses it.
- Even if user authentication is enabled, the authentication mechanism may be weak or easily circumvented.
- Once authentication is enabled, changing the authentication information regularly is seldom done.
- Limit processing power of the device, may preclude the use of computationally intensive authentication techniques or cryptographic algorithms.

Smart card authentication is perhaps the best-known example of a proof by possession mechanism. Other classes of authentication mechanisms include proof by knowledge (e.g., passwords) and proof by property (e.g., fingerprints). Smart cards are credit-card-size, plastic cards that host an embedded computer chip containing an operating system, programs, and data, and can be imprinted with a photo, a magnetic strip, or other information, for dual use as a physical identification badge [Pol97]. When used for user authentication, smart cards can help to improve the security of a device as well as provide additional security services. Many organizational security infrastructures incorporate smart cards. However, standard size smart cards are generally not amenable to handheld devices because of the relatively large size of the card, the need for a suitable card reader, and the difficulty and cumbersomeness of interfacing a reader to the device.

This report provides an overview of two novel types of smart card that use standard interfaces supported by most handheld devices, instead of those interfaces favored by most smart card readers. The report describes how these forms of smart card can be used for authenticating users on handheld devices and provides details of the solutions' design and implementation. The authentication mechanisms were implemented in C and C++ on an iPAQ Personal Digital Assistant (PDA), running the Familiar distribution of the Linux operating system from handhelds.org and the Open Palmtop Integrated Environment (OPIE). OPIE is an open-source implementation of the Qtopia graphical environment of TrollTech. OPIE and Qtopia are both built with Qt/Embedded, a C++ toolkit for graphical user interface (GUI) and application development for embedded devices, which includes its own windowing system. The Familiar distribution was modified with a multi-mode authentication framework [Jan03a] that includes a policy enforcement engine, which governs the behavior of both code modules and users [Jan03b]. That framework provides the facility to add new authentication mechanism modules and have them execute in a prescribed order.

Background

Smart cards are designed to protect the information they contain, and usually require a PIN to verify the user's identity before granting access to on-card information. The computer chip on the card requires a smart card reader to obtain power and a clock signal and to communicate with the computing platform. Tamper resistance techniques are used to protect the contents of the chip. Once contact is made with the reader, a smart card uses a serial interface to communicate with software running on the computing platform. Java Card is currently one of the more popular operating systems for smart cards. Data units received by the smart card are processed by Java applications installed on the card. The Java Card runtime facilitates the development and deployment of Java applications to support authentication and other security-related applications, such as those involving electronic commerce.

PDA/Smart Card Interfaces

The capabilities and form factor of smart cards are compatible with some handheld devices, provided that a reader can somehow interface with the device and a compatible driver is available for the platform's operating system. For example, a number of manufacturers produce smart card readers as hardware modules that fit into a type II PCMCIA Card slot. These readers accept standard size smart cards, which can be obtained separately. A platform, such as an iPAQ 5550 PDA, whose expansion options include both single and double PCMCIA slot expansion sleeves, can readily accept such readers and operate them, once a suitable driver is found and installed. More elegant solutions also exist such as the Blue Jacket from Axcess Mobile Communications, which incorporates a smart card reader within the expansion sleeve and can support an optional Bluetooth communications and a type II compact flash interface [Blue]. In either case, however, such solutions are limited to certain types of PDAs and add considerable bulk to the device.

Smart cards come in other form factors. A popular format emerging for smart cards is a USB key fob. This chewing-gum-pack sized hardware component has a printed circuit board with a processor and memory encased within a plastic housing with a USB connector at one end. Many manufacturers produce USB devices that function identically to smart cards and, since they interface through a USB port, eliminate the need for a reader. Currently, however, very few PDA devices support host USB ports, which are needed to interface to these peripherals. One constraining factor is that the PDA would need to draw on its battery to power any peripherals plugged into the USB port.

Another alternative is the iButton, a 16mm computer chip contained in a stainless steel shell, able to be mounted in jewelry (e.g., ring, bracelet, necklace) [Dal02]. Capabilities of these button size devices range from a simple memory token to a microprocessor and arithmetic accelerator able to support a Java Card-compliant Virtual Machine. However, a button receptacle incorporated into the device or easily added (e.g., via a compact flash card) is needed to facilitate their use with PDA devices. USB holders are also available for iButtons, but require a host USB port. Thus, one obstacle in using smart card functionality to authenticate users on handheld devices is packaging the functionality in a form factor that is compact, unencumbering, and compatible with the capabilities possessed by handheld devices.

PC/SC Protocol Stack

In addition to establishing a suitable hardware interface, a software protocol stack is needed for a device to communicate with a smart card. One of the best-known general-purpose architectures for supporting smart cards on desktop systems is the PC/SC (Personal Computer/Smart Card) specification [PCSC]. Though originally aimed at Windows based systems, the specification has evolved and been adapted for other platforms, including Linux. The MUSCLE group created the open source PC/SC software stack for Linux, named PC/SC Lite, which also runs on Linux-based PDAs.[1] Figure 1 illustrates the organization of the PC/SC Lite software with respect to a smart card and reader.

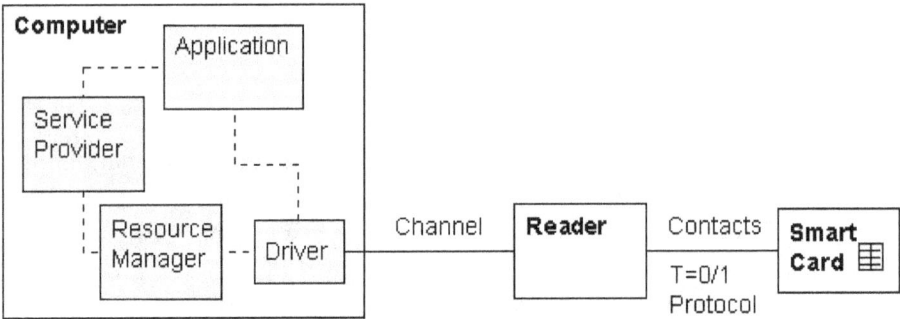

Figure 1: PC/SC Lite

Three main software components support a smart card application: the service provider, a resource manager, and a driver for the smart card reader. The software application normally communicates with the driver indirectly via the service provider, which in turn uses the standardized PC/SC interface to the resource manager. Similarly, the resource manager uses a standard interface to communicate with the driver. The driver is closely tied to characteristics of the smart card reader, while the service provider is closely tied to the characteristics of the smart card. This functional separation allows an application to use any smart card with any reader, provided that the respective service provider and driver software are available at the host computer. Getting the correct service provider and driver software activated is the job of the resource manager. Note that, if needed, the application can bypass these PC/SC components and communicate directly with the driver, loosing the aforementioned benefits of the scheme.

User Authentication with Smart Cards

For a smart card to allow access, it typically requires the user to input a PIN (Personal Identity Number) first, to verify that the individual in possession of the card is the person to whom the card was issued. Incorrect PINs keep the card from functioning and eventually cause it to lock. Once the PIN is successfully entered, a dialogue between the PDA and smart card occurs, by which the PDA confirms that the card and the credentials of the user on the card are valid and correspond to that of the PDA owner. The underlying mechanism used to authenticate users via smart cards relies on a challenge-response protocol between the device and the smart card. The PDA challenges the smart card for an appropriate and correct response that can be used to verify that the card is the one originally enrolled by the device owner. The PDA relies on user

---

[1] More information can be found at http://www.linuxnet.com/

4

credential information, obtained earlier from the smart card when the PDA owner initially enrolled the card with the device.

The authentication mechanisms discussed in this report use a challenge-response protocol compliant with FIPS 196 to authenticate a user to the PDA. Figure 2 illustrates a typical exchange, omitting the requisite PIN satisfaction step that occurs. The upper part of the diagram shows the enrollment information exchange used to register a card (at right) with the PDA (at left), while the remainder show the exchanges used to verify the claimed identity.

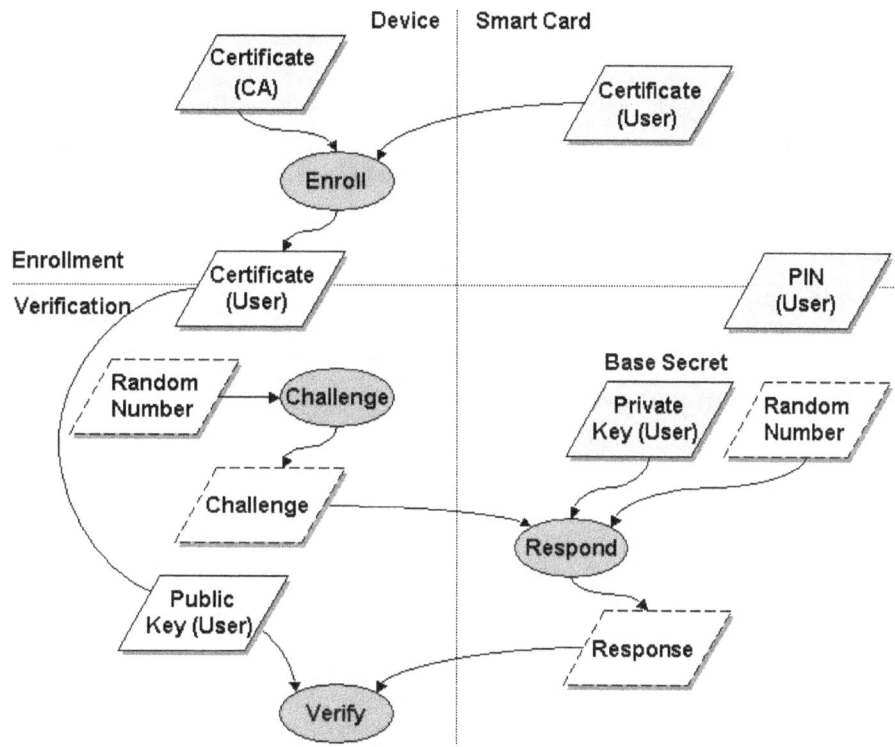

Figure 2: Challenge-Response Exchange

Before the smart card can be enrolled at the device, it must first be personalized for the user. The personalization step is essentially another enrollment process whereby a security administrator enrolls the user on the card (i.e., user enrollment), by populating it with the user's credentials, PIN, and other information. Those credentials are obtained from the card when it is enrolled with the device (i.e., card enrollment) and validated using the certificate of the Certification Authority (CA) who issued the credentials or is otherwise the point of trust for validation. The issuing of credentials from a CA is described later in Appendix C of this report. Once the credentials are validated, they are retained at the device and used to verify the user's identity following FIPS 196 procedures.

For verification, the device and smart card adhere to the following unilateral entity authentication protocol [FIPS196]:

- The device, acting as the verifier, generates a random challenge "B" and passes it to the smart card for signing with the private key associated with the enrolled identity certificate;
- The smart card, acting as the claimant, generates a random value "A", signs A∥B with the private key on the card ('∥' denotes concatenation), and returns A and the signature to the PDA;
- The device retrieves the enrolled identity certificate, verifies it, then verifies the card's signature over A∥B using the public key in the certificate
- If everything successfully verifies, authentication succeeds; otherwise, the authentication attempt fails

The authentication of an entity depends on two things: the verification of the claimant's binding with its key pair, and the verification of the claimant's digital signature on the random number challenge. Using a private key to generate digital signatures for authentication makes it computationally infeasible for an attacker to masquerade as another entity, while using random number challenges prevents an intruder from copying a valid response signed by the claimant and replaying it successfully at a later time. Including a random number of the claimant in the response before signing it precludes the claimant from signing data that is solely defined by the verifier. The security of the FIPS 196 protocol also hinges on the generation of random numbers that have a low probability of being repeated.

The Multi-mode Authentication Framework (MAF)

MAF was developed previously in a related effort to provide a structured environment for the protection and execution of one or more authentication mechanisms operating on Linux handheld devices [Jan03a]. The authentication mechanisms described in this report were implemented specifically for this framework. Each authentication mechanism consists of two parts: an authentication handler and a user interface (UI). Figure 3 illustrates these elements within a Linux operating system environment, enhanced with kernel support for MAF.

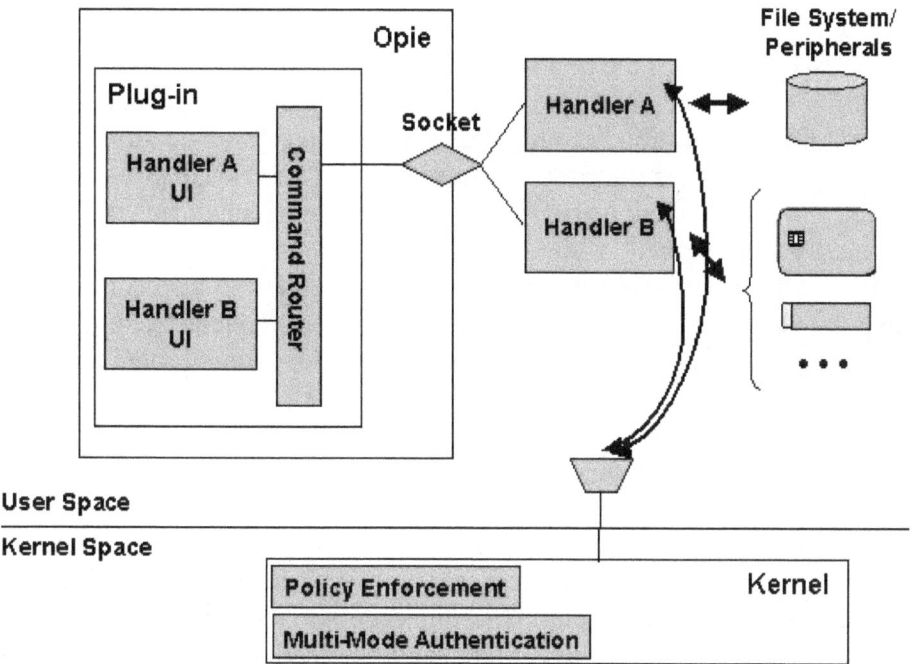

Figure 3: Multi-mode Authentication Framework

Authentication handlers embody the procedure that performs the actual authentication. They communicate with the kernel, listening for when to initiate authentication and reporting whether authentication was successful. They communicate with the user interface components to bring up specific screens, accept input, display messages, etc. on the device. Handlers also communicate with any peripheral hardware devices needed for authentication, such as a security token, and access the file system to store and retrieve information as needed. Handlers run in user space as do their respective user interface.

The user interface for an authentication mechanism is implemented as a set of components of a plug-in module for OPIE. Their function is to perform all necessary interactions with the user. For example, with smart card applications they can be used to prompt for and accept a PIN entry or to notify the user of errors. The plug-in module supports a socket interface to receive commands from an authentication handler that runs as a separate process, and to route the commands to the correct user interface component. Similarly, reverse routing is also supported for responses from user interface components to an authentication handler.

The kernel has two key modifications to support the framework: the multi-mode authentication functionality and the policy enforcement functionality.

7

- Policy enforcement's main responsibility is to impose different sets of policy rules on the device, as signaled by multi-mode authentication, for one or more defined policy contexts referred to as policy levels. For example, it can block hardware buttons and certain I/O ports on the device until the user is authenticated at the lowest policy level, policy level 1. Policy enforcement is also used to protect authentication information files, the user interface and handler components, and policy enforcement information against improper access. Moreover, it also has the means to register and start up authorized handlers, if they are not running, or restart them, if they terminate for some reason.

- The main responsibility of the multi-mode authentication functionality within the kernel is to govern the authentication steps as they relate to the various policy levels that are configured. Communication between the kernel and an authentication handler is done via the /proc file system. The multi-mode authentication functionality maintains complete knowledge about the mappings between authentication mechanisms and policy levels, simplifying the development of the authentication handlers. One of its key functions is to initiate user authentication when the device is powered on. It also controls the order and frequency in which the handlers are awakened from suspended state and begin execution, and ensures that messages from only legitimate handlers are accepted and processed.

Together, the kernel policy enforcement and multi-mode authentication extensions are essential for securing authentication applications.

To create an additional authentication mechanism, a developer needs to create a new authentication handler along with any required user interface objects and the policy rules to protect the mechanism. Policy rules include limiting access to the storage objects used, the user interface objects within the plug-in module, and the authentication handler itself. They also can limit communications to peripheral devices and among the handler, the user interface, and kernel. Note that writing an authentication mechanism that neither interacts with the user nor requires a user interface component is possible. For example, the mechanism could be based on a sensor that is continually monitored and whose input triggers both an authenticated or non-authenticated transition.

Smart Multi-Media Card Authentication

The Smart Multi-Media Card (SMMC) authentication mechanism relies on a smart card chip packaged in a multimedia card format in lieu of a traditional size smart card. The postage stamp-size card houses an MMC controller, smart card, and additional flash memory. Many PDAs support a combination secure digital (SD)/MMC card slot, making such cards a viable means to provide smart card functionality. The MultiMediaCard Association has recently drafted standard specifications for secure multimedia cards.

A pre-production Smart MMC produced by Renesas called the X-Mobile Card (XMC) observes the draft standard and was used for the prototype implementation.[2] The tamper resistant hardware module complies with Java Card 2.2.1, Global Platform 2.1, FIPS 140-2 (currently under evaluation), and other standards. To use the X-Mobile Card, a Linux device driver had to be developed. The resulting Smart MMC driver is available at an open source site that specializes exclusively in smart card software for Linux.[3]

Overview

The SMMC handler monitors card insertion and removal, and controls all the necessary steps regarding the authentication mechanism. The handler communicates with a special purpose on-card applet called "Enroller" that validates the PIN and uses the FIPS 196 challenge-response protocol to verify the user's claimed identity. The applet and user's PIN are placed on the smart card along with other information during card personalization.

The SMMC handler, as all MAF handlers, runs in user space and communicates with the Linux kernel and with the OPIE plug-in components that make up the user interface. The handler also uses the PC/SC Lite layer to talk to the XMC smart card and the "Enroller" Java Card applet placed on the card during personalization.

In communicating with the kernel, the handler tells the kernel that it is a polling handler and specifies the polling interval. It receives orders from the kernel either to poll the smart card or to perform authentication. The handler replies to the kernel giving either the result of the authentication (success or failure) or, whenever the card state changes to "present," an indication for the kernel to call it back with an order to perform authentication so that an attempt to raise the security level to this level can be made.

In the communication with the OPIE plug-in, the handler tells the UI to display certain informative messages, when needed, or to accept PIN entry from the user.

The communication with the card consists of exchanging APDUs (application protocol data units) with the Java Card applet resident on the card over a secure channel. The aim is twofold: to authenticate the user to the PDA when asked by the kernel, and to ensure that the card is the same one as at the previous polling moment. Interactions with the card require that the correct PIN, set during card personalization, be first supplied. At the first authentication attempt, the

---

[2] More information can be found at http://www.x-mobilecard.com/products/product.jsp
[3] More information can be found at http://www.linuxnet.com

card is registered with the PDA; the user's public key certificate is retrieved from the card onto the PDA and verified using the root certificate of the issuing CA.

User authentication follows the FIPS 196-compliant challenge/response protocol described earlier. The handler verifies the signature using the user's public key contained in the user's public key certificate that was downloaded onto the PDA at registration.

Protection

For user authentication the fundamental threat is an attacker impersonating a user and gaining control of the device and its contents. Smart cards are designed to resist tampering and monitoring of the card, including sophisticated attacks that involve reverse engineering, fault injection, and signal leakage. Presuming those designs are effective, the following vulnerabilities are the main candidates for exploitation:

- The authentication mechanism can be bypassed
- Weak authentication algorithms and methods are used
- The implementation of a correct and effective authentication mechanism design is flawed
- The confidentiality and integrity of stored authentication information is not preserved

The SMMC mechanism relies on MAF, which in turns relies on the security of the underlying operating system implementation. The handler is protected from being substituted or overwritten with another program through the multi-mode authentication and policy enforcement functionalities of MAF. Substitution is prevented through an entry in the list of registered handlers (</usr/bin/handlerSMMC 2>) identifying its location, while overwrite is prevented using the following policy rule in the MAF policy file (/etc/MAF/defaultPolicy) [Jan03b], which also grants exclusive access to the file containing the certificate of the user maintained by the handler once retrieved from the smart card:

<file /root/Settings/SMMCcert.pem /usr/bin/handlerSMMC 0>

The SMMC handler uses the following data files stored on the PDA, which must also be protected through policy enforcement functionality of MAF:

- /etc/MAF/SMMCauthkey: contains the 16-byte authentication key used to set up the secure channel of communication to the XMC card.
- /etc/MAF/SMMCmackey: contains the 16-byte MAC key used to set up the secure channel of communication to the XMC card.
- /etc/MAF/cacert.pem: contains the X.509 root certificate of the CA (whose role is carried out by the Card Enroller application)

The following policy rules grant exclusive permission to read/write these files to the handler at any policy level:

- <file /etc/MAF/cacert.pem /usr/bin/handlerSMMC 0>
- <file /etc/MAF/SMMCauthkey /usr/bin/handlerSMMC 0>
- <file /etc/MAF/SMMCmackey /usr/bin/handlerSMMC 0>

10

The SMMC token builds on the protection afforded by the hardware and requires a correct PIN to unlock its functions. Several bad PIN entry attempts lock the card. A Global Platform Secure Channel is used to protect the PIN and any other information sent from the device to the card. The private key of the user and the user's PIN established during personalization cannot be exported from the token.

Handler Implementation

The SMMC handler operates as a polling handler, periodically checking the status of the smart card as well as initiating authentication with it. The following code excerpt shows the main execution loop of the handler.

```
crtCardState = CARD_ABSENT;
while (1)
{
    prevCardState = crtCardState;
    result = HandlerReady(3);
    if (result == mmPoll)
    {
        crtCardState = GetCardState();
        if (crtCardState == CARD PRESENT &&
            crtCardState == prevCardState)
        {
            if (CardWasReinserted()) crtCardState = CARD ABSENT;
        }
        if (crtCardState != prevCardState)
        {
            if (crtCardState == CARD_ABSENT)
            {
                TellKernel("AUTH-FAIL");
            }
            else
            {
                TellKernel("LEVEL 1");
            }
        }
    }
    else
    {
        TellKernel(Login()? (char*)"AUTH-OK" : (char*)"AUTH-FAIL");
    }
}
```

The variable crtCardState stores the card's current state as reported by the GetCardState() function. The state "absent" (CARD_ABSENT) means the card is removed from the card reader or unusable; the state "present" (CARD_PRESENT) means the card is inserted in the reader. GetCardState() uses the PC/SC Lite API ScardGetStatusChange() to obtain the card state. The variable prevCardState maintains the card state as detected at the previous polling moment, allowing a change in the card state from one polling moment to the next to be detected.

After saving the previous card state in prevCardState, the handler tells the kernel that it is a polling handler by calling the HandlerReady() function with the polling interval of 3 seconds.

11

The kernel directs the handler to either poll the card (i.e., the HandlerReady result is mmPoll) or to perform an authentication procedure (the HandlerReady result is mmAuthenticate). Performing authentication is accomplished with a call to the Login() function and returning the verdict. A successful Login() establishes an authentication session with the smart card, which involves obtaining the PIN from the user, then establishing the secure channel with the card and issuing a challenge and verifying the response using the appropriate APDUs defined in Appendix A.

```
BOOL Login()
{
    char msgfromUI[100];

    UIAttach();

    // If a card is already inserted, try to authenticate the user
    // before displaying the prompt "Insert card" on the interface.
    // UserIsAuth returns 0 = success, 1 = failure, 2 = cancel.
    if ((crtCardState = GetCardState()) == CARD PRESENT) {
        switch (UserIsAuth())
        {
            case 0: return TRUE;
            case 1: Flash("Wrong card!"); return FALSE;
            case 2: return FALSE;
        }
    }

    // If card is absent or unusable, make UI display prompt.
    TellUI("SMMC:shw:Please insert card!");

    // Loop waiting for the user to insert the card.
    while ((crtCardState = GetCardState()) == CARD ABSENT)
    {
        // See if we have something from UI.
        if (PollUI())
        {
            RecvUI(msgfromUI, sizeof(msgfromUI));
            // If we got "Cancel", clear UI and return failure.
            if (strcmp(msgfromUI + 5, "Cancel") == 0)
            {
                TellUI("SMMC:clr:");
                return FALSE;
            }
        }
        else
        {
            sleep(1);
        }
    }

    // Now card is inserted.
    TellUI("SMMC:clr:");

    // Try to authenticate the user.
    Switch (UserIsAuth())
    {
```

```
      case 0: TellUI("SMMC:clr:"); return TRUE;
      case 1: Flash("Wrong card!"); return FALSE;
      case 2: return FALSE;
    }
    // Shouldn't reach this point, but...
    return FALSE;
  }
```

Performing the polling operations is more complicated. Polling operations begin with obtaining the card's current state by calling GetCardState() at the top of the main loop. The handler must take into consideration whether a previous authentication session exists, the card's current and previous states, and whether the card was reinserted (and maybe replaced) between the previous and the current polling moments. The possible cases that can occur during a poll and the action taken by the handler in each case appear in Table 1.

Table 1: Decision Matrix

Previous Authentication Session Exists	Card's Current State	Card's Previous State	Card Reinserted	Action
Yes	Present	Present	Yes	Tell kernel the authentication failed
			No	Do nothing
		Absent	N/A	Tell kernel that it may try to raise the policy level
	Absent	Present	N/A	Tell kernel the authentication failed
		Absent	N/A	Do nothing
No	Present	Present	No	Do nothing
		Absent	N/A	Tell kernel that it may try to raise the policy level
	Absent	Present	N/A	Tell kernel the authentication failed
		Absent	N/A	Do nothing

The handler detects whether the card was removed and inserted (or a different one inserted) by calling the function CardWasReinserted(). Removing and inserting the card requires starting a new authentication session, with the first APDU of the new session being a SelectApplet. Therefore, if no previous authentication session exists, the card is not considered "reinserted." If a previous authentication session exists, however, the function tries to send an APDU (which is not a SelectApplet nor preceded by a SelectApplet) to the card. If the card was removed and inserted (or a different one inserted), the APDU fails because a SelectApplet is required for the new session.

The most interesting case in the table is the first entry, when the card appears present, but was removed and inserted again (and possibly replaced). To force reauthentication in this situation, the handler changes the current card state to "absent" (see the source code), while the previous state remains "present." The next few lines of code tell the kernel that the authentication failed, but in the next iteration the previous card state is updated from the current card state and current

13

card state is refreshed, resulting in prevCardState and crtCardState to be set respectively to "absent" and "present." Subsequently, the handler tells the kernel that it may raise the level, so that in the next iteration the HandlerReady result is mmAuthenticate, triggering card authentication by calling the Login() function.

Token Implementation

The Enroller applet is was designed for use in Renesas' X-Mobile Cards (XMC) and other Java Card-compliant smart cards. Its name is a bit of a misnomer, since the applet participates both in the personalization of the card and the authentication process. The applet conforms to Java Card 2.1.2 specifications and supports secure channel communications between a host application and itself as specified in Global Platform 2.1. The Enroller applet has the following functionality:

- Supports the optional creation of a secure channel of communication between the host application and the applet.
- Generates an RSA private/public key pair (1024 bits).[4]
- Stores an RSA public or private key (1024 bits) onto the XMC card.
- Retrieves the RSA public key previously stored on the XMC card.
- Stores an X.509 certificate as a byte array on the XMC card.
- Retrieves the X.509 certificate previously stored on the XMC card.
- Sets user PIN.
- Verifies user PIN.
- Replies to a host challenge by signing it with the private RSA key previously stored on the XMC card.

The specific Application Protocol Data Units (APDUs) supported by the Enroller applet are defined in Appendix A.

The remainder of this section describes in more detail two important applet operations, involving setting up the secure channel and responding to a challenge request.

Setting up the Secure Channel

To set up a secure channel with the smart card, the host application sends the Initialize Update and External Authentication commands to the Enroller Applet. The Enroller Applet processes the commands by simply passing them onto the default security domain, the CardManager applet, for processing.

Each time the Enroller applet is selected, it first retrieves a handle to an object that implements the SecureChannel interface:

---

[4] Note that the private key needs to be used to sign a certificate request. Because generating a certificate request on the card is rather complicated, the keys and the certificate request are generated on the enrollment station, and the keys transferred to the card.

```
package enroller;

import javacard.framework.*;
import javacard.security.*;
import javacardx.crypto.*;
import org.globalplatform.*;

public class Enroller extends javacard.framework.Applet {

    private SecureChannel mySecChan;

    public boolean select() {
        ...
        mySecChan = GPSystem.getSecureChannel();
    }
```

Then the Enroller applet uses the security domain's processSecurity() method for that handle to process the Initialize Update and External Authenticate APDUs and establish the secure channel:

```
private void processSecApdu(APDU apdu) {
    short len = (short)0;
    try {
        len = mySecChan.processSecurity(apdu);
    } catch (…) {
        ...
    }
    if (len != 0) {
        apdu.setOutgoingAnd Send((short)ISO7816.OFFSET CDATA, len);
    }
}
```

From then on, each APDU sent through the secure channel by the handler must have the class bit identified by the mask 0x04 set on (e.g., 0x94 instead of 0x90) and the data field encrypted, as prescribed by the Global Platform documentation. The Enroller applet must decrypt the data field of incoming APDUs. To avoid unnecessary code duplication, the Enroller applet calls the unwrap() method of the default security domain, as shown below, to perform this function:

```
private void processApdu(APDU apdu) {
    short dataLen = apdu.setIncomingAndReceive();
    byte[] buf = apdu.getBuffer();
    mySecChan.unwrap(buf, (short)0, (short)(dataLen + 5));
    /* process decrypted apdu*/
    ...
}
```

Signing the host challenge

To authenticate the smart card/user, the handler generates a random value of 16 bytes, known as randomB, and sends it to the card in the SignChallenge APDU. When the applet receives the challenge it must prepare a response to be signed with the user's private key maintained on the card. The Enroller applet generates its own secure random value, known as randomA, and replies with randomA plus the signature over randomA || randomB, as shown in the code below.

15

```
byte[] retBuf = new byte[256];
byte[] digest = new byte[128];
byte[] randomA = new byte[CHALLENGE_LENGTH];

MessageDigest md = MessageDigest.getInstance(MessageDigest.ALG_SHA,
false);
Cipher rsaCipher = Cipher.getInstance(Cipher.ALG_RSA_PKCS1, false);

void signChallenge(APDU apdu) {
    unwrapApdu(apdu);
    byte[] apduBuf = apdu.getBuffer();

    boolean simpleAuth = ((apduBuf[ISO7816.OFFSET_P1] & 0xFF) == 0x00);
    challengeSigned = false;

    if (pin == null || !pin.isValidated() || !privateKeyLoaded) {
        ISOException.throwIt(ISO7816.SW_CONDITIONS_NOT_SATISFIED);
    }
    if (simpleAuth && (userDn == null)) {
        ISOException.throwIt(ISO7816.SW_CONDITIONS_NOT_SATISFIED);
    }

    short ranBLen = (short)(apduBuf[ISO7816.OFFSET_LC] & 0xFF);
    if (ranBLen <= 0) {
        ISOException.throwIt(ISO7816.SW_WRONG_LENGTH);
    }

    // Generate randomA in return buffer starting at offset 2
    // (first 2 bytes are reserved for total length and randomA length)
    RandomData rData =
RandomData.getInstance(RandomData.ALG_SECURE_RANDOM);
    retBuf[1] = CHALLENGE_LENGTH;
    rData.generateData(retBuf, (short)2, CHALLENGE_LENGTH);

    // Digest randomA, randomB.
    md.reset();
    md.update(retBuf, (short)2, CHALLENGE_LENGTH);
    short digLen = md.doFinal(apduBuf, (short)ISO7816.OFFSET_CDATA,
ranBLen,
        digest, (short)0);

    // Encrypt the digest. Put the signature length and the signature
    // in retBuf after randomA.
    rsaCipher.init(userPrivateRSAKey, Cipher.MODE_ENCRYPT);
    short sigLen = rsaCipher.doFinal(digest, (short)0, (short)digLen,
retBuf,
                (short)(3 + CHALLENGE_LENGTH));
    retBuf[(short)(2 + CHALLENGE_LENGTH)] = (byte)sigLen;
    short totLen = (short)(2 + CHALLENGE_LENGTH + sigLen);
    retBuf[0] = (byte)(totLen);

    // Set flag indicating the challenge was signed.
    challengeSigned = true;

    // Send response.
```

```
        apdu.setOutgoing();
        apdu.setOutgoingLength((short)(1 + totLen));
        apdu.sendBytesLong(retBuf, (short)0, (short)(1 + totLen));
}
```

All the cryptographic objects and buffers used in signing the challenge are statically allocated; otherwise, repeating this operation (the applet will be used by a polling application) eventually exhausts the available memory, since garbage collection is not available on the card.

Bluetooth Smart Card (BSC) Authentication

The Bluetooth Smart Card (BSC) authentication mechanism relies on a smart card chip packaged together with other components in a compact-size form factor (e.g., a key fob or Bluetooth-enabled cell phone) in lieu of a traditional size smart card. Rather than bringing a smart card into physical contact with a PDA, as with the SMMC, a wireless interface is used. Bluetooth Smart Card authentication can provide the security of smart card-based authentication to a PDA with the following advantages:

- No need exists for a specialized smart card reader on the PDA
- The token can be small enough to fit comfortably in a pocket or a purse
- It can work within a few meters of the PDA, even without a direct line of sight
- It does not require much power, nor need to draw power from the PDA
- It can be discrete (i.e., non-discoverable by third parties)

A Bluetooth Smart Card token in the form factor of a key fob would house a Bluetooth radio, smart card, processor and memory, and battery. The token could also include an LCD to allow PIN entry and management directly, independently of another device. Since many PDAs support a Bluetooth radio, such tokens are a viable means to bring smart card functionality to a device. A BSC token could also be used with Bluetooth-enabled workstations. The mechanism is also amenable to other types of low-power Personal Area Network (PAN) communications, should they become popular and integrated into PDAs. Conceivably, a token that supports more than one type of PAN communications is possible.

Overview

The Bluetooth Smart Card has many similarities with the SMMC insofar as both solutions depend on the functionality of a Java Card-compliant smart card chip, use the same challenge-response protocol for user authentication, and are implemented to execute within the MAF environment. Therefore, wherever possible, components of the SMMC were reused for BSC.

The main difference from SMMC is that communications between the PDA and token takes place using a Bluetooth channel rather than an MMC bus. Another difference is that PIN entry may occur at the BSC token rather than at the PDA.

In developing the solution, an effective way was found to split the PC/SC functionality between the PDA and BSC token to allow Bluetooth communications, yet have minimal impact on any smart card application. Figure 4 illustrates where the PC/SC Lite components were divided and allocated between the PDA and BSC token. The small box appearing between the resource manager and driver, called the IFD handler, proved to be the key.[5] The IFD Handler provides a standard interface to the resource manager on one side and maps the functions over the Bluetooth channel to the other, permitting the BSC token to implement an entire IFD subsystem independently of the other PC/SC Lite components. APDUs are sent over the L2CAP Bluetooth layer using a socket interface.

---

[5] The IFD handler is a PC/SC Lite component and unrelated to MAF handlers except by the unfortunate similarity of name.

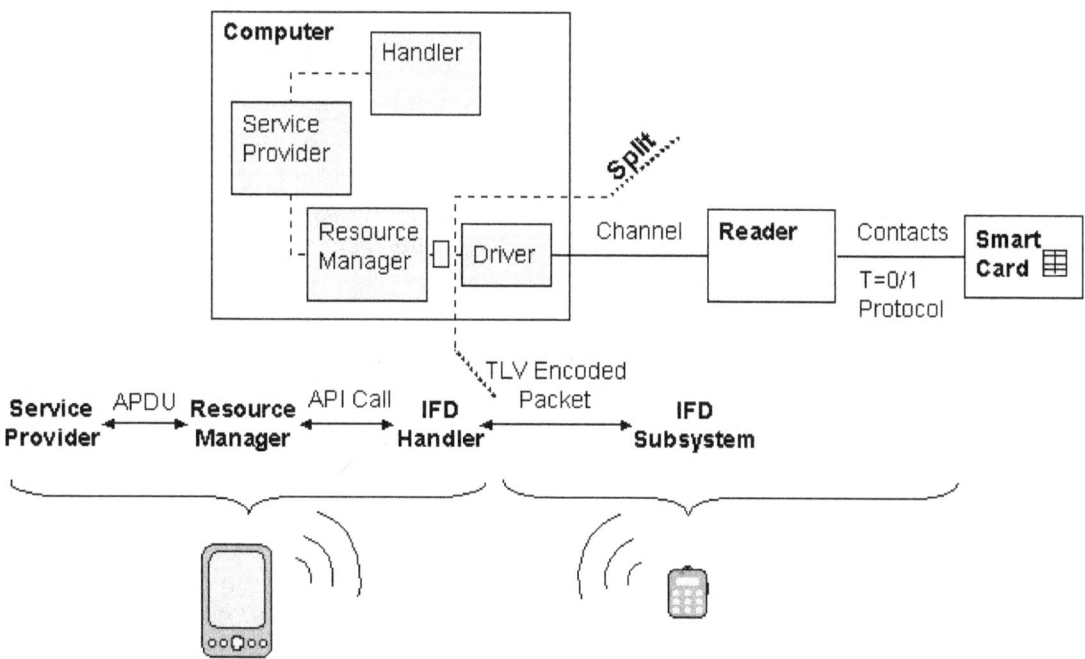

Figure 4: PC/SC Lite Reification

The arrangement allows the Bluetooth Smart Card solution to work with any type of Java Card-compliant smart card recognized by the Movement for the Use of Smart Cards in a Linux Environment[6] and the PC/SC Lite framework.[7] For simplicity, however, X-Mobile Cards were used in the token prototype, running the same Enroller applet used with the SMMC.

The BSC handler functions nearly identically to the SMMC handler. It runs in user space and communicates with other framework components and needed peripherals, including the Linux kernel, the OPIE user interface plug-in components, and the XMC smart card personalized with the same "Enroller" Java Card applet used for the SMMC. Because the IFD handler for the BSC token provides the identical interface as that of the IFD handler for the SMMC token, the BSC handler can, in principle, use either type of token interchangeably.

Protection

Recall that for user authentication the fundamental threat is an attacker impersonating a user and gaining control of the device and its contents. Assuming smart cards designs result in secure products, the following vulnerabilities as with the Smart MMC token are the main candidates for exploitation:

- The authentication mechanism can be bypassed
- Weak authentication algorithms and methods are used
- The implementation of a correct and effective authentication mechanism design is flawed
- The confidentiality and integrity of stored authentication information is not preserved

---

[6] More information can be found at http://www.linuxnet.com/
[7] More information can be found at http://alioth.debian.org/projects/pcsclite/

The BSC mechanism on the PDA relies on MAF, which in turn relies on the security of the underlying operating system implementation. As with SMMC solution, the handler is protected from being substituted and overwritten respectively through the multi-mode authentication and policy enforcement functionalities of MAF. Because of the similarity of the two implementations, ignoring for the moment the IFD interface differences, same policy rules used for SMMC apply to BSC as well. Similarly, the BSC token builds on the same protection features afforded by the smart card in the SMMC solution for its security.

Communications Concerns

Because the BSC solution uses Bluetooth, a wireless radio-based communication channel, the solution must address the following issues not present in the SMMC or traditional solutions involving a wired client-reader-smart card paradigm:

- An attacker can eavesdrop on the communication channel from a distance
- An attacker can send information to the device and PDA via an active Bluetooth interface to attempt to impersonate one or both parties, or to disrupt communications
- When the PDA selects a token with which to communicate, the PDA cannot be certain it contacted its user's token
- When a token is contacted, the token cannot be certain the device that contacted it is the device of its user

Communications Security

Two Bluetooth devices should be paired to one another during the initial enrollment phase, when the token is first registered with the PDA. Bluetooth pairing establishes a shared 128-bit symmetric key on both devices. The BlueZ Unix Bluetooth stack used in the implementation appends the key to the binary file /etc/bluetooth/link_keys. The common symmetric key is then used to authenticate (via a MAC on Bluetooth L2CAP packets) and encrypt (via a specific stream cipher) exchanged packets. In addition, the device address of the paired communication partner is retained to eliminate discovery processing. The following policy rules are needed to protect the device address and the Bluetooth symmetric key on the device:

- <file   /etc/MAF/BSCmac   /usr/bin/handlerSMMC   0>
- <file   /etc/bluetooth/link_keys   /usr/bin/handlerSMMC   0>
- <file   /etc/bluetooth/link_keys   /usr/sbin/hcid 0>[8]

At the application level, pre-established symmetric keys (i.e., one for MAC, and one for TripleDES encryption) are also used to protect transmitted APDUs using the Global Platform Secure Channel (GPSC) Protocol 01 format. Under this protocol, only the data portion of an APDU is encrypted and encryption is applied unidirectionally from the PDA to the smart card. If needed, the host and card applications could be modified to encrypt all or parts of the messages returned from the smart card, but this is not currently part of the GPSC Protocol. However, aside from the PIN transmission, none of the other information exchanged (i.e., the FIPS 196 challenge and response) affects the overall mechanism if exposed.

---

[8] This rule lets Bluetooth use the link key when it connects, since the previous rule prevents any other library from reading the keys.

Figure 5 illustrates the two levels of confidentiality protection afforded by GPSC and Bluetooth. The shaded portions represent the encrypted parts of the dialogue.

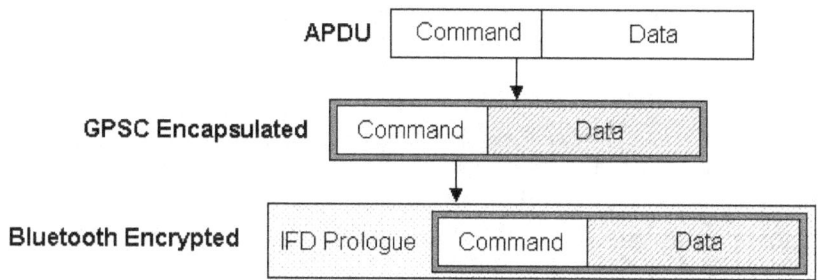

Figure 5: Communication Protection

It is possible to disable non-authenticated communications on both the PDA and token immediately after enrollment. After that, the device can only be contacted if its MAC address is known, at least in theory. If another pairing operation is disallowed, authentication becomes mandatory and quite secure, since the token talk exclusively to the PDA and vice-versa. This is an additional, but minimal barrier, because some tools (e.g., bluesniff, btscanner, and redfang) can discover devices when they are set to be non-discoverable. Moreover, Bluetooth security is not as strong as RSA and FIPS 196-based authentication, especially if the users (or the processes) do not use extremely long PINs to pair the devices and/or attackers are present during the initial pairing (and link key establishment) operation.

Handler Implementation

As mentioned previously, the SMMC handler on the PDA side as well as the SMMC applet on the card side were mostly reusable in the BSC implementation. Only a small part of the SMMC handler needed to be augmented to enable remote PIN verification. Remote PIN verification is a request from the handler to the IFD subsystem on the token to inquire about its capabilities to accept PIN entries. If the capability is not present, the handler prompts the user for this information on the PDA as done for the SMMC. If the capability is present, the handler bypasses those steps and instead relies on the BSC token to obtain the PIN from the user. This change works equally well for the SMMC and BSC tokens, allowing the same handler to be used for both authentication mechanisms. Thus, the part of the BSC implementation that distinguishes it from the SMMC is the IFD handler developed to communicate over Bluetooth to the IFD subsystem on the token.

The IFD handler is software executing on the PDA that implements a standard, hardware-independent, and I/O channel-independent interface into the IFD subsystem. The job of IFD handler is to map the standard interface it offers onto the Interface Device functionality (e.g., a smart card reader), so that data can be exchanged with a smart card. Communicating with a device driver often suffices. However, in this case, since the BSC token supports a complete IFD subsystem independently from the host, the IFD handler performs the interface mapping to the smart card functionality through communications over a Bluetooth channel.

For the Bluetooth Smart Card a new IFD handler, called bt_ifd, was implemented. The IFD handler acts as an interface for a normal Smart Card reader driver for the PC/SC Lite stack, but it operates as a proxy for another IFD present on the BSC token which is reached via Bluetooth.

21

Three versions of the standardized IFD interface definition exist. Since most smart card reader drivers follow the version 2 specifications, an IFD version 2 interface was implemented for the BSC IFD Handler. On the token side of the Bluetooth channel, virtually any reader supporting the same IFD version can be employed, in addition to the Smart MultiMedia Card driver used in the implemention. For example, this architecture was tested with another reader and Java Card-compliant smart card, a GemXpresso 211/PK card with a PCMCIA GPR400 GemPlus reader with positive results. The PCSC lite stack on the host recognized the GemPlus reader and the card in it, as if they still were connected locally to the smart card client. Thus, this division allows the implementation of the IFD subsystem to take advantage of standard device drivers, I/O channels, and IFD Handlers, available for various smart card chips.

The protocol used to forward the IFD functions and arguments to the BSC token, and to receive the corresponding responses, is a custom Tag-Length-Value-based serialization protocol. Any forward message starts with a byte representing the IFD function being transmitted:

```
#define CREATECHANNEL        1
#define CLOSECHANNEL         2
#define GETCAPABILITIES      3
#define SETCAPABILITIES      4
#define SETPROTOCOLPARAMETERS    5
#define POWERICC             6
#define TRANSMITTOICC        7
#define CONTROL              8
#define ICCPRESENCE          9
```

From this identifier, both the client and the server know the number of the arguments, their type, and their order for each defined function. Following the identifier byte, comes each argument one after the other, in proper sequence, encoded as follows:

- If the argument is fixed length, it is directly appended. Only two types of fixed arguments are supported: DWORD, 4 bytes long, and UCHAR, 1 byte long.
- If the argument is variable length, it is preceded by its length and then directly appended. The length value is encoded as a DWORD of 4 bytes, sufficient to represent any size parameter envisioned.

The return messages are even simpler, since most of them are only a 4 bytes long RESPONSECODE (like a DWORD). When arguments are returned with the RESPONSECODE, they are serialized one by one using the same encoding as used for the forward messages, as described before. Examples of the serialization protocol are given in Appendix D.

Token Implementation

The token was implemented as an OPIE application running on an iPAQ PDA (which can be a StrongARM PDA. e.g. h3870, as well as an XScale one, e.g., an h3970 or h5550). A virtual token appears on the PDA's display and works as the real token would. Figure 6 shows a screen shot of a token in a key fob form factor that approximates its intended size and with appropriate components (Bluetooth chip, ARM processor, small battery and embedded smart card). Note that the core of this application does not depend on any OPIE specific features such as Bluetooth support. The server that performs the token's functions runs as a C binary, as is described later in this section. The GUI interface uses Qt signals and slots for communication and some graphical widgets for display, but the latter would not be necessary on a real token.

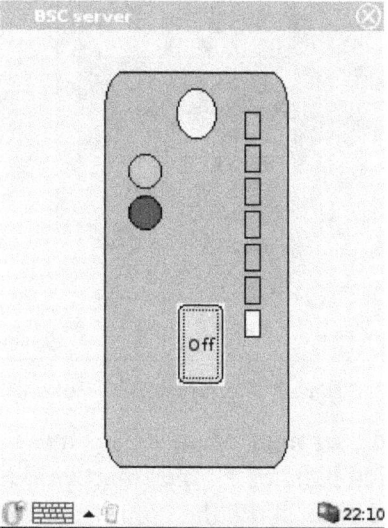

Figure 6: Bluetooth Smart Card Token

A user interacts with the token through the On/Off button. By default, neither Bluetooth nor the BSC (Bluetooth Smart Card) server are launched, but starts when the On/Off button is pressed (the text 'Off' appears in place of 'On', when the latter is pressed and vice versa). The order of activities is as follows:

- Bluetooth is started, which is indicated by the lower circular LED at left becoming blue.
- The BSC server (a simple mono-threaded socket server forwarding messages to the IFD interface of the SMC smart card) is started, which is indicated by the top circular LED at left becoming green.
- The server is then operational, and represents its activity in real time through the stacked LEDs at right, showing the quantity of output lines the BSC server receives within a fixed time.

Another version of the token allows the user to enter the smart card PIN locally on an LCD display, completely avoiding any Bluetooth eavesdropping attack. Figure 7 shows the virtual token displayed on the PDA screen. The same common functions as the previous variant are supported. However, adding a numeric pad with cancel, clear and OK buttons, and a pseudo-LCD screen that can display messages allows PIN entry and even passkey entry for Bluetooth pairing (see Appendix E).

23

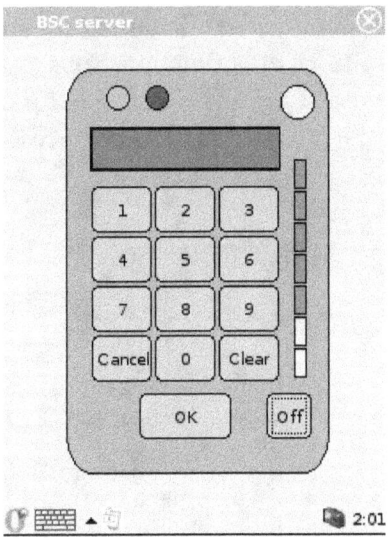

Figure 7: PIN Entry BSC Token

As explained earlier, the graphical user interface (GUI) is mainly a wrapper around a command-line binary core, written in C, which handles all the smart card and Bluetooth socket interactions. The GUI, which is an OPIE application, starts the command line binary server (called ifd_server) when the "On" button is pressed. The same server is used by both tokens (with and without a PIN keypad). A flag (--nopin) tells the binary if it should run in a remote PIN verification mode or not. Once the server is active, the LED indicator is lit and the GUI then reacts to the output from the server. Any output triggers the activity LEDs at right in proportion to quantity of output.

The GUI application for the PIN keypad detects, through a special output message, when the server needs a PIN, and reacts by displaying a "Please enter PIN" message on the LCD screen and activating the PIN keypad. When the user eventually enters a PIN and presses OK or Cancel, the GUI sends back a special command to the server, to inform it either that the PIN was entered (and its value), or that it was cancelled.

This organization of the server makes it simple to use a different smart card reader, since the two types of GUI interfaces and the command-line server are independent from any IFD driver (as long as it implements the IFD standard interface version 2). The command-line binary, ifd_server, is built linking the server code, ifd_server.c, with the object file containing the IFD driver being used (e.g. smc_ifd.o or gpr400_ifd.o). Therefore, by just putting the new smart card reader driver with the ifd_server.c source file, and building them together, results in a BSC server that supports a different smart card reader.

Card Personalization

Card personalization involves populating information such as the user's name and credentials into files maintained on the smart card, before issuing to a user. In addition, the procedure usually entails recording a PIN on the card, which the user later enters to enable the card to confirm that the holder is the same person who was issued the card. The personalization procedure can also involve physical preparation of the card, such as imprinting pictures, names, and address information on the surface. In the case of programmable cards, additional programs and data files can be installed to enable specific functions for a user.

Personalization of the Secure MMC token and Bluetooth Smart Card token is performed by an Enrollment Station. The Enrollment Station is a Windows platform that runs the CardEnroller application. The CardEnroller application performs the following tasks when personalizing a token:

- Loads a Javacard applet, named Enroller, onto the XMC card
- Sets the user PIN as selected by the user
- Verifies the user PIN (required in subsequent operations)
- Generates a pair of RSA keys and obtains a corresponding X.509 certificate for the user.
- Stores the private RSA key and the certificate containing the public key on the card
- Destroys its copy of the user private key

The CardEnroller application also performs the functions of a Certificate Authority when obtaining a certificate, generating and signing the certificate itself. Its (root) certificate is exported to any PDAs needing to verify a user's credentials that it issues. In a production environment, this function would be performed by a Certificate Authority operating within a Public Key infrastructure for the organization.

An installation guide and user's guide for the Enroller application are given respectively in Appendices B and C.

Summary

While mobile handheld devices provide productivity benefits, they also pose new risks associated with the information and network access capabilities they accumulate over time. User authentication safeguards against the risk of unauthorized use and access to a device's contents. This paper demonstrates how novel forms of smart cards can be implemented as a primary authentication method suitable for mobile devices. The methods used take advantage of interfaces built into most handheld devices, affording organizations the opportunity to extend smart card functionality easily to a segment of the computational infrastructure here to fore neglected.

References

[Blue]   Blue Jacket Product Information, Axcess Mobile Communications, <URL: http://www.axcess-mobile.com/products/BlueJacketFlyer.pdf>.

[Dal02]   iButton Overview, Maxim/Dallas Semiconductor Corp, 2002, <URL: http://www.ibutton.com/ibuttons/index.html>.

[FIPS196]   Entity Authentication Using Public Key Cryptography, Federal Information Processing Standards Publication (FIPS PUB) 196, U.S. Department of Commerce, National Institute of Standards and Technology, February 18, 1997, <URL: http://csrc.nist.gov/publications/fips/fips196/fips196.pdf>

[Jan03a]   Wayne Jansen Vlad Korolev, Serban Gavrila, Thomas Heute, Clément Séveillac, A Framework for Multi-Mode Authentication: Overview and Implementation Guide, NISTIR 7046, August 2003, <URL: http://csrc.nist.gov/publications/nistir/nistir-7046.pdf>.

[Jan03b]   Wayne Jansen, Tom Karygiannis, Michaela Iorga, Serban Gavrila, Vlad Korolev, Security Policy Management for Handheld Devices, The 2003 International Conference on Security and Management (SAM'03), June 2003, <URL: http://csrc.nist.gov/mobilesecurity/Publications/SecurityPolicyManagementForPDAs-IEEEformat.pdf>.

[Pol97]   Despina Polemi, Biometric Techniques: Review and Evaluation of Biometric Techniques for Identification and Authentication, Institute of Communication and Computer Systems, National Technical University of Athens, April 1997, <URL: ftp://ftp.cordis.lu/pub/infosec/docs/biomet.doc>.

[PCSC]   Interoperability Specification for ICCs and Personal Computer Systems, Part 1. Introduction and Architecture Overview, Revision 2.00.11, PC/SC Workgroup, May 2004, <URL: http://www.pcscworkgroup.com/specifications/files/pcsc1_v20.pdf>.

Appendix A - Supported Application Protocol Data Units (APDUs)

An APDU is a packet of information representing either an instruction sent to the card or the response from the card after executing the instruction. The Enroller applet supports specific APDUs of class 0x90 (or 0x94 if the secure channel mode of communication is enabled). In addition, the applet supports the APDUs Initialize Update and External Authenticate, which are used to establish the secure channel of communication.

In the APDU descriptions that follow, all numbers are in hexadecimal unless otherwise specified. The bytes of an APDU are indicated using the following conventions:
- CLA – class of instruction,
- INS – instruction,
- P1 and P2 – parameters 1 and 2,
- Lc – length of the command data,
- Data – command or response data,
- Le – length of the expected response, and
- SW1 and SW2 – status words 1 and 2.

Generate RSA Private/Public Key Pair

Generates a pair of 1024_{10} bits RSA keys using the javacard.security.KeyPair class. Stores the keys (on the card) as persistent objects.

CLA	INS	P1	P2	Lc
90 or 94	10	00	00	0

CLA: 90 without secure channel communication, 94 with secure channel communication.

Response:

SW1	SW2

SW1	SW2	Description
90	00	No error.
69	86	Command not allowed (applet is in APPLET_BLOCKED state).
69	85	Conditions not satisfied (PIN wasn't set and/or verified prior to this command).

Set Private RSA Key

Stores a private RSA key on the card. Actually, two APDUs are necessary to store the key: one for the modulus and one for the exponent. The only key size supported is 1024_{10} bits.

CLA	INS	P1	P2	Lc	Data
90 or 94	20	00	xx	Data length	Data

CLA: 90 without secure channel communication, 94 with secure channel communication.
P2: identifies the key component that is set:
    00: the key modulus;
    01: the key exponent.
Data: the key component as a string of 128_{10} bytes (1024_{10} bits).

Example: 90 20 00 01 80 2D 34 EF... sets the exponent of the private key to the string of 0x80 bytes 2D 34 EF...

Response:

SW1	SW2

SW1	SW2	Description
90	00	No error.
69	86	Command not allowed (applet is in APPLET_BLOCKED state).
69	85	Conditions not satisfied (PIN wasn't set and/or verified prior to this command).
6A	86	Incorrect P2 (neither 00 nor 01).
67	00	Wrong data length (not 0x80 = 128).

Set Public RSA Key

Stores a public RSA key on the card. Actually, two APDUs are necessary to store the key: one for the modulus and one for the exponent. The only key size supported is 1024_{10} bits.

CLA	INS	P1	P2	Lc	Data
90 or 94	22	00	xx	Data length	Data

CLA: 90 without secure channel communication, 94 with secure channel communication.
P2: identifies the key component that is set:
    00: the key modulus;
    01: the key exponent.
Data: the key component as a string of 3-to-128_{10} bytes.

Example: 90 22 00 01 03 01 00 01 sets the public key exponent to 01 00 01.

Response:

SW1	SW2

SW1	SW2	Description
90	00	No error.
69	86	Command not allowed (applet is in APPLET_BLOCKED state).
69	85	Conditions not satisfied (PIN wasn't set and/or verified prior to this command).
6A	86	Incorrect P2 (neither 00 nor 01).
67	00	Wrong data length (not 80 for modulus or between 03 and 80 for exponent.

Get Public RSA Key

Retrieves the public RSA key from the card, where it was stored using the "Set public RSA key" APDU. Two APDUs are necessary to retrieve the key: one for the modulus and one for the exponent. The only key size supported is 1024_{10} bits. The modulus will always be 128_{10} bytes; the exponent may be from 3 to 128_{10} bytes.

CLA	INS	P1	P2	Data length
90 or 94	26	00	xx	00

CLA: 90 without secure channel communication, 94 with secure channel communication.
P2: identifies the key component that is retrieved from the card:
    00: the key modulus;
    01: the key exponent.

Example: 90 26 00 01 00 requests the public key exponent.

Response:

Data	SW1	SW2

Data: contains the requested component of the key, preceded by a byte containing its length. For the modulus, the length should be 80. For the exponent, the length may be between 03 and 80. Example of response: 03 01 00 01 90 00.

SW1	SW2	Description
90	00	No error.
69	86	Command not allowed (applet is in APPLET_BLOCKED state).

69	85	Conditions not satisfied (PIN wasn't set and/or verified, or the public key wasn't set prior to this command).
6A	86	Incorrect P2 (neither 00 nor 01).

Set Certificate Length

Sets the length of and allocates the byte array needed to store a X.509 certificate.

CLA	INS	P1	P2	Lc	Data
90 or 94	28	00	00	02	Data

CLA: 90 without secure channel communication, 94 with secure channel communication.
Data: the certificate length on 2 bytes.

Example: 90 28 00 08 02 05 5F tells the applet to allocate 55F bytes for the certificate that will be stored on the card.

Response:

SW1	SW2

SW1	SW2	Description
90	00	No error.
69	86	Command not allowed (applet is in APPLET_BLOCKED state).
69	85	Conditions not satisfied (PIN wasn't set and/or verified prior to this command).
67	00	Wrong data length (Lc is not 2).

Set Certificate Data

Appends a block of certificate data to the certificate portion previously stored on the card. The applet keeps track of the current offset in the certificate byte array. The offset is initialized to zero when the certificate length is set.

CLA	INS	P1	P2	Lc	Data
90 or 94	2A	00	00	Data length	The certificate data block

CLA: 90 without secure channel communication, 94 with secure channel communication.
Data length: the length of the certificate block on one byte.
Data: the certificate data block. Its length must be less than or equal to 80.

Example: 90 2A 00 00 05 2F 4D 33 51 6D appends 5 bytes (2F 4D 33 51 6D) to the portion of certificate already uploaded to the card.

Response:

SW1	SW2

SW1	SW2	Description
90	00	No error.
69	86	Command not allowed (applet is in APPLET_BLOCKED state).
69	85	Conditions not satisfied (PIN wasn't set and/or verified, or certificate length wasn't set prior to this command).
63	D2	Data block too long (the block doesn't fit into the certificate byte array based on its length and current offset).

Get Certificate Length

Obtains the certificate length from the card.

CLA	INS	P1	P2	Data length
90 or 94	2C	00	00	00

CLA: 90 without secure channel communication, 94 with secure channel communication.

Response:

Data	SW1	SW2

Data: the length of the certificate on 2 bytes.

SW1	SW2	Description
90	00	No error.
69	86	Command not allowed (applet is in APPLET_BLOCKED state).
69	85	Conditions not satisfied (PIN wasn't set and/or verified, or certificate length wasn't set prior to this command).

Get Certificate Data

Obtains a certificate block from the card. The offset of the block in the certificate is specified in P1 and P2. The maximum length of the certificate block is specified in one byte of data and must be less than or equal to 80 (actually if it's greater than 80, the applet sets it to 80). The applet tries to return a block of the maximum length specified. If the specified maximum length exceeds the length of the remaining certificate portion starting at the specified offset, then the applet returns the remaining certificate portion.

CLA	INS	P1	P2	Lc	Data
90 or 94	2E	Offset in certificate		Data length	Data

CLA: 90 without secure channel communication, 94 with secure channel communication.
P1 and P2: the offset (in the certificate's byte array) of the certificate block to be returned.
Data length: 1.
Data: the maximum length of the certificate block to be returned, on 1 byte.

Example: 90 2E 01 00 01 80 requests a certificate block starting at offset 100 in the certificate array and having a maximum length of 80.

Response:

Data	SW1	SW2

Data: the requested certificate block.

SW1	SW2	Description
90	00	No error.
69	86	Command not allowed (applet is in APPLET_BLOCKED state).
69	85	Conditions not satisfied (PIN wasn't set and/or verified, or certificate length wasn't set, or certificate data was not set prior to this command).
67	00	Wrong data length (if data length is not 1).
6A	86	Incorrect P1, P2 (the offset is beyond the end of the certificate byte array).

Set User PIN

Sets the user PIN to a sequence of 4 to 8 decimal digits and the maximum number of tries to 3.

CLA	INS	P1	P2	Lc	Data
90 or 94	34	00	00	Data length	Data

CLA: 90 without secure channel communication, 94 with secure channel communication.
Data length: the length of the PIN (4 to 8).
Data: the PIN (4 to 8 decimal digits).

Example: 90 34 00 00 05 31 32 33 34 35 sets the user PIN to "12345".

Response:

SW1	SW2

SW1	SW2	Description
90	00	No error.
69	86	Command not allowed (applet is in APPLET_BLOCKED state).
69	85	Conditions not satisfied (PIN is already set and was not verified prior to this command).
67	00	Wrong data length (if data length is not between 4 and 8).

Verify User PIN

Verifies the user PIN.

CLA	INS	P1	P2	Lc	Data
90 or 94	32	00	00	Data length	Data

CLA: 90 without secure channel communication, 94 with secure channel communication.
Data length: the length of the PIN (4 to 8).
Data: the PIN (4 to 8 decimal digits).

Example: 90 32 00 00 05 31 32 33 34 35 verifies whether the user PIN is "12345".

Response:

SW1	SW2

SW1	SW2	Description
90	00	No error.
69	86	Command not allowed (applet is in APPLET_BLOCKED state).
69	85	Conditions not satisfied (PIN was not set prior to this command).
63	Cx	Pin not verified; x tries remain.

Set User Id

Set the user identifier (known as userId) on the card. The user identifier is supposed to uniquely identify the user; it may be the user's distinguished name, for example.

CLA	INS	P1	P2	Lc	Data
90 or 94	30	00	00	Data length	Data

CLA: 90 without secure channel communication, 94 with secure channel communication.
Data length: the length of the userId.
Data: userId.

Response:

SW1	SW2

SW1	SW2	Description
90	00	No error.
69	86	Command not allowed (applet is in APPLET_BLOCKED state).
69	85	Conditions not satisfied (PIN not yet set or not yet verified.

Sign Challenge

Signs the host-generated challenge and a card-generated random value, and returns the card-generated random value and the signature. The applet uses the SHA1 algorithm for digest and the RSA private key for encryption.

CLA	INS	P1	P2	Lc	Data
90 or 94	38	Xx	00	Data length	Data

CLA: 90 without secure channel communication, 94 with secure channel communication.
P1: xx is 00 for the simple token authentication, 01 for the certificate-based authentication.
Data length: the length of the host challenge.
Data: the host-generated challenge (known as randomB).

Response:

Data	SW1	SW2

Data:

For the *simple token authentication*, the returned data consists of:
1 byte: total length of the returned data (excluding SW1, SW2).
1 byte: length of the card-generated random value, known as randomA.
The card-generated random value randomA.
1 byte: length of signature.
The signature over (userId || randomA || randomB). UserId is the user identifier, previously set on the card.
1 byte: length of userId.
The user identifier userId.

For the *certificate-based token authentication*, the returned data consists of:
1 byte: total length of the returned data (excluding SW1, SW2);
1 byte: length of the card-generated random value, known as randomA;
The card-generated random value randomA;
1 byte: length of signature;
The signature over (randomA || randomB).

SW1	SW2	Description
90	00	No error.
69	86	Command not allowed (applet is in APPLET_BLOCKED state).
69	85	Conditions not satisfied (PIN not yet set or not yet verified, and/or the private RSA key not yet set). For simple authentication, userId not yet set.

Initialize Update

Initialize Update is the first step (card authentication) in setting up a secure channel of communication between the host application and the card.

CLA	INS	P1	P2	Lc	Data
80	50	00	00	Data length	Data

Data length: the length of the host challenge for the session.
Data: the host challenge (host session data), as prescribed by Global Platform. It is made up of 8 bytes of 00.

Response:

Data	SW1	SW2

Data: 1C bytes containing card data valid for the session.

SW1	SW2	Description
90	00	No error.

67	00	Incorrect length.
6A	86	Incorrect P1, P2

External Authenticate

External Authenticate is the second step (host authentication) in setting up a secure channel of communication between the host application and the card.

CLA	INS	P1	P2	Lc	Data
84	82	Security control parameter	00	Data length	Data

P1: 0x03: MAC plus encryption; 0x01: MAC; 0x00: mutual authentication.

Response:

SW1	SW2

SW1	SW2	Description
90	00	No error.
67	00	Incorrect length.
6A	86	Incorrect P1, P2
63	00	Cryptogram not verified.
69	85	Command conditions not satisfied.
6A	88	MAC not verified.

Appendix B – CardEnroller Installation Guide

Hardware and software requirements

The following items are needed to install and run the CardEnroller application:

- A desktop or laptop computer running Windows 98/2000/XP.
- A XMC smart card reader.
- The executable CardEnroller.exe.
- The OpenSSL DLL libeay32.dll.
- The CardEnroller's (root) X.509 certificate file and private key file.
- The authentication and MAC keys (provided by Renesas) of the XMC card for establishing a secure channel between CardEnroller and the card.
- The converted applet file (CAP file) for the Javacard Enroller applet.
- The applet identifiers (AIDs) for the XMC card manager applet, the Enroller applet package, and the Enroller applet instance.

Preparing the Card Enroller keys and certificate

OpenSSL is used to illustrate this step, but any tool that generates a pair of RSA keys and a self-signed X.509 certificate in PEM format can be used.

Step 1. Prepare a configuration file (a text file), with the following contents:

```
[req]
default_bits            =       2048
default_keyfile         =       C:\\CardEnroller\\cakey.pem
default_md              =       sha1

prompt                  =       no
distinguished_name      =       root_ca_distinguished_name

x509_extensions         =       root_ca_extensions

[root_ca_distinguished_name]
emailAddress            =       serban.gavrila@nist.gov
countryName             =       US
stateOrProvinceName     =       Maryland
localityName            =       Gaithersburg
organizationName        =       NIST
organizationalUnitName=         CSD
commonName              =       XMC Enroller

[root_ca_extensions]
basicConstraints        =       CA:true
```

The values of default_bits, default_keyfile, default_md, emailAddress, countryName, stateOrProvinceName, localityName, organizationName, organizationalUnitName, commonName can be edited to match a particular site's characteristics (all these are attributes of the certificate issuer – the CardEnroller in our case, and not of the client requesting a certificate). The default_keyfile value is the complete path of the file where the CardEnroller's private key will be stored encrypted with a password.

Step 2. Assuming that the path of the above configuration file is C:\CardEnroller\openssl.conf, the following commands should be run next:

C:\>cd CardEnroller
C:\CardEnroller>set OPENSSL_CONF=C:\CardEnroller\openssl.conf
C:\CardEnroller>openssl req –x509 –newkey rsa –out cacert.pem –outform PEM

The CardEnroller certificate will be placed in the file C:\CardEnroller\cacert.pem, and the CardEnroller's private RSA key in the file C:\CardEnroller\cakey.pem, encrypted with the password provided when prompted by the command openssl. The password must be retained, because it is required later when the CardEnroller application signs the issued certificate(s). Openssl uses the PEM format for both key and certificate, as instructed.

Installing the Card Enroller

Copy the OpenSSL DLL libeay32.dll to a folder in the main path. Copy all other needed files to their appropriate location(s), consistent with the configuration file.

Create a configuration file %USERPROFILE%\certenroller.conf as detailed below. %USERPROFILE% is an environment variable in all Windows versions. To see its value, type "echo %USERPROFILE%" in a DOS shell.

The configuration file is a text file, which contains a set of properties of the form key=value, one property per line. Empty lines or comments (lines beginning with a '#' character) are allowed and ignored. The configuration file must be present at the above-specified location, and it must define properties for the following keys (capitalization is important):

- capFname: the path of the CAP file for the Javacard Enroller applet.
- cardManAid: the AID of the Card Manager applet on the XMC card.
- packageAid: the AID of the Javacard Enroller applet package.
- instanceAid: the AID of the Javacard Enroller applet instance.
- authKey: the key used for mutual authentication between the host and the card.
- macKey: the key used for data encryption in the secure communication between the host and the card.
- cacertFname: the path of the file containing the Card Enroller's certificate.
- cakeyFname: the path of a file containing the Card Enroller's private key encrypted with a password. The Card Enroller will ask for the password interactively.
- serialFname: the path of a file containing the next serial number to be used in a generated certificate. If the file is missing, the Card Enroller will use 1 as next serial number and will create a serial file with the appropriate contents.
- secureChannel: "yes" or "no", depending on whether a Global Platform secure channel is to be established between the off-card application and the Javacard applet.

The following is an example of configuration file:

```
capFname="C:\\CardEnroller\\enroller\\javacard\\enroller.cap"
cardManAid=A0000001510000
packageAid=B000000001
instanceAid=B00000000101
authKey=404142434445464748494A4B4C4D4E4F
macKey=404142434445464748494A4B4C4D4E4F
cacertFname="C:\\CardEnroller\\cacert.pem"
cakeyFname="C:\\CardEnroller\\cakey.pem"
serialFname="C:\\CardEnroller\\serial"
secureChannel=yes
```

Note: the property values specified in the configuration file can be overwritten by using CardEnroller's menu "Tools/Configure…".

Appendix C – CardEnroller Operation Guide

Begin by launching CardEnroller.exe. If the configuration file %USERPROFILE%\certenroller.conf does not exist, or a required configuration property is not defined, the execution is aborted. Otherwise, CardEnroller reads the configuration file and displays its main window on the screen, as shown in Figure 8.

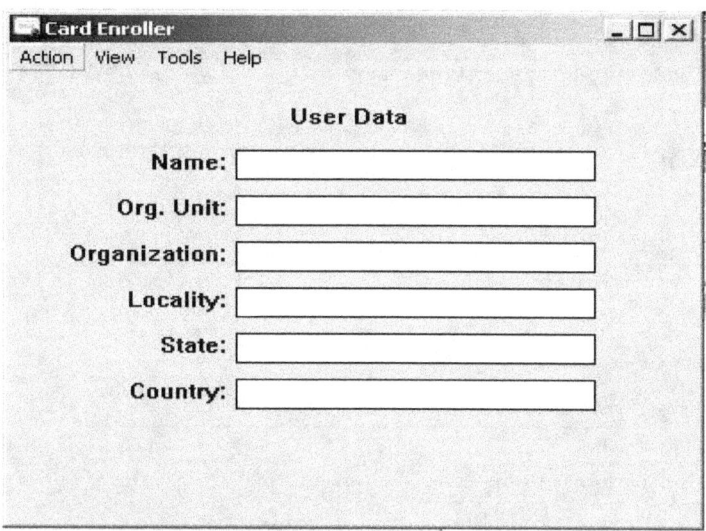

Figure 8: Card Enroller's Main Window

To check or modify the configuration properties, select the menu "Tools/Configure …." The configuration window shown in Figure 9 appears, displaying the current configuration properties and allowing their values to be edited. The (new) values are saved in the configuration file when the OK button is clicked.

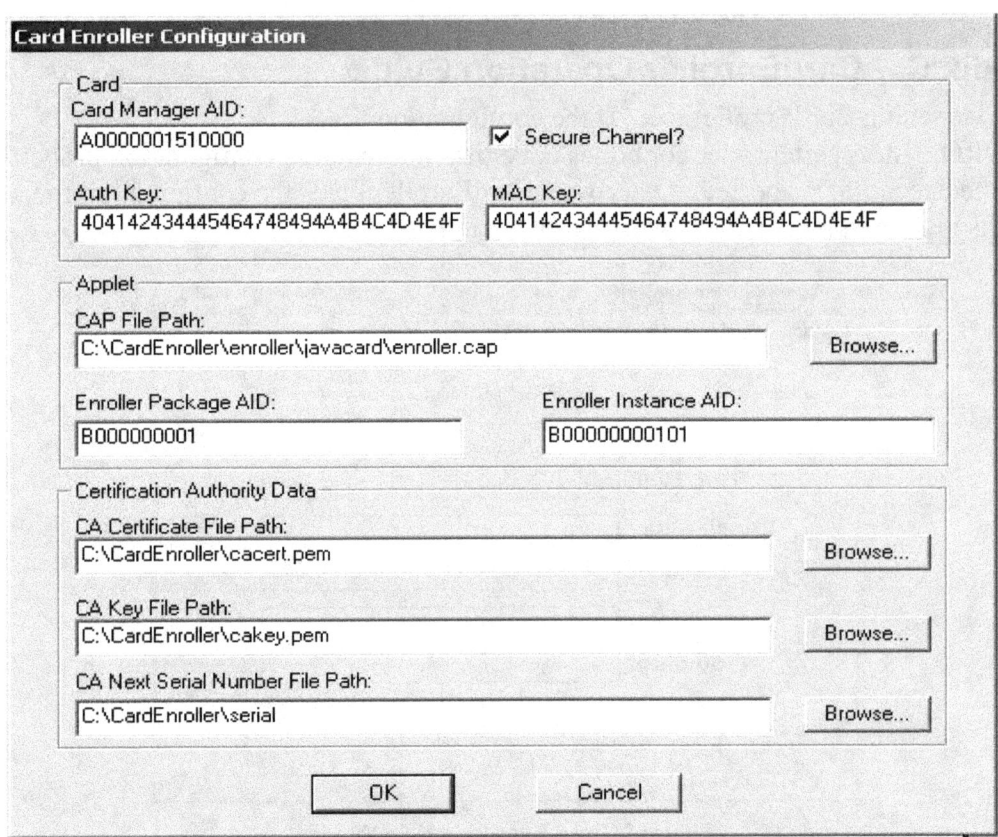

Figure 9: Card Enroller Configuration Window

To personalize a XMC card for a user, the user data in the fields of the main window must be entered, as illustrated in Figure 10. Insert the XMC card in the card reader, and then select the menu "Action/Enroll."

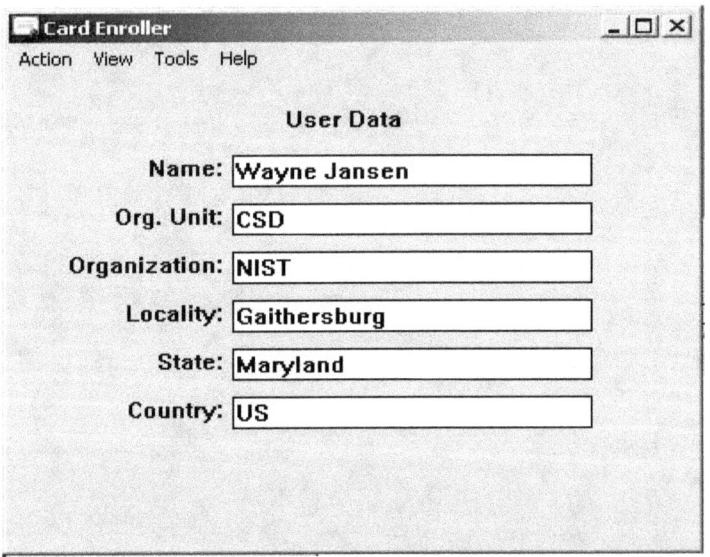

Figure 10: Main Window with User Credentials

Note that, at a minimum, the user name is needed to enroll. The CardEnroller tool first establishes a secure channel of communication with the card, then performs the following steps:

42

- Loads the Enroll applet onto the card (deleting possible old applet instances).
- Asks the user for a PIN (4-8 decimal digits) and set the user PIN.
- Verifies the user PIN just set (it will ask again the user for the PIN).
- Generates a pair of RSA keys and a X.509 certificate for the user and stores the private key and the certificate onto the card. Note that to sign the certificate, the CardEnroller application needs its private key, which was stored in a file encrypted with a password, as described in Appendix B, "Preparing CardEnroller keys and certificate", Step 2. You must provide the same password when prompted by CardEnroller.

The lower part of the CardEnroller's main window displays the steps as they are performed.

The other menus of the CardEnroller application can be used to:

- Display the last certificate generated by the CardEnroller tool or read from the smart card (the menu "View/Last Certificate"). For an example, see Figure 11.
- Extract and display the certificate from a XMC card (the menu "View/Card Certificate").
- Generate a new RSA key pair and certificate (the menu "Action/Get Certificate"). Note that Action/Enroll also generates new user keys and certificate (so the old ones will be lost) that will be used to personalize the smart card.
- Save the last certificate generated by the CardEnroller tool or read from the smart card to a file in PEM format by using the menu "Action/Save Certificate....". If the certificate was generated by the tool together with the keys and not simply read from the smart card, then the associated private key will also be saved encrypted with a password if you check the menu "Action/Save Private Key".

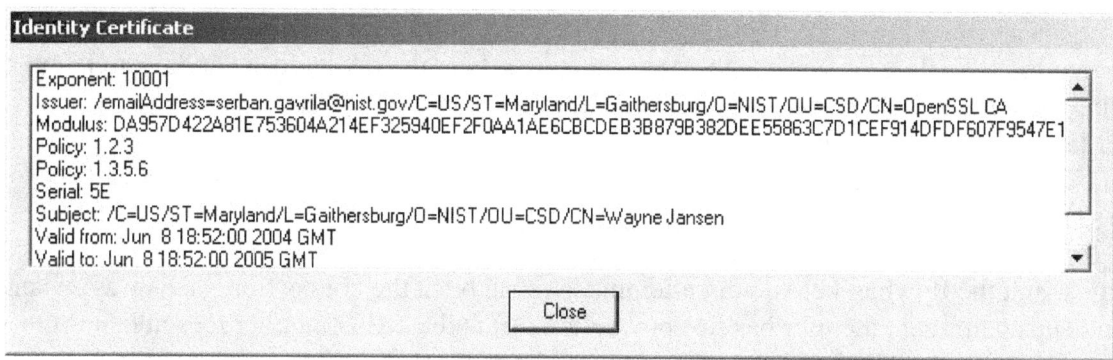

Figure 11: A Displayed Certificate

Appendix D - IFD Handler Interface Examples

Below are listed examples of key IFD functions, their arguments and purpose, and the translation of typical argument values into bytes serialized for transmission and reception on the Bluetooth link between the client and the token.

IFDHICCPresence ([in] DWORD Lun)

This function returns the status of the card inserted in the reader/slot specified by Lun.

Transmission: 09 00 00 00 00

- 09 is the ICCPRESENCE function code (all the defined functions codes are one byte long and listed earlier in the handler implementation section of the BSC mechanism).
- 00 00 00 00 is a DWORD (i.e., four bytes long or four pairs of hex chars) equal to zero Lun is zero, since the first slot is always used.

Reception: 09 67 02 00 00

- 09 is the ICCPRESENCE function code.

- 67 02 00 00 is the RESPONSECODE (a DWORD), equal to IFD_ICC_PRESENT (615 = 0x00000267 in hex, in Little Endian mode where the least significant byte comes first – used by Familiar on ARM iPAQs, and on all x86 Desktop PCs, if a card is in the reader). Otherwise IFD_ICC_NOT_PRESENT (616 = 0x00000268 in hex, 68 03 00 00 on the network) is returned as the DWORD value.

Note that the IFDHICCPresence is the only function where the token does not have to be up and running for a reply to be made (one could say that is the only function where we lie a little to the upper layers). Indeed, when the token is off or unreachable, which can happen quite frequently, (e.g., if the user has forgotten the token at home, has powered off the token to save battery power or to simply prevent access, etc.), IFD_RESPONSE_TIMEOUT or IFD_COMMUNICATION_ERROR is not returned. Instead, IFD_ICC_NOT_PRESENT is returned, and the IFD handler silently attempts to reestablish the connection. This way, when the token is up again (e.g., the user has just powered it on) and the IFD handler reestablishes the connection, IFD_ICC_PRESENT can be returned to the next PS/SC resource manager inquiry. This is transparent to well-designed smart card applications, like the SMMC/BSC handler, because they continually poll the PC/SC Lite stack when there is no card (through the SCardGetStatusChange function, which calls the IFDHICCPresence one). Hence, they know when the card status changes and they can start sending commands for the card again.

IFDHGetCapabilities ([in] DWORD Lun, [in] DWORD Tag, [in,out] PDWORD Length, [out] PUCHAR Value)

This function should return the slot/card capabilities for a particular slot/card specified by Lun. The tag of the desired capability, Tag, and the maximum expected length (in bytes), Length are also indicated. The response code plus the return value Value, of length Length are returned.

Transmission: 03 00 00 00 00 02 00 00 42 01 00 00 00

- 03 is the GETCAPABILITIES function code (defined earlier)

- 00 00 00 00 is a DWORD value for Lun equal to zero, the first slot is always used.
- 02 00 00 42 is a DWORD value for Tag, equal to IOCTL_SMARTCARD_VENDOR_VERIFY_PIN (not yet defined in PCSClite main API, but suggested in the IFD official documentation and in a PC/SC Lite developer email).

- 01 00 00 00 is a DWORD value for Length, equal to 1 here. Indeed, 1-byte answer is expected to a VERIFY_PIN GetCapabilities question.

Reception: 03 00 00 00 00 01 00 00 00 01

- 03 is the GETCAPABILITIES function code

- 00 00 00 00 is the RESPONSECODE equal to IFD_SUCCESS (0x00000000) if the capability described by the Tag exists, or IFD_ERROR_TAG (0x00000600) otherwise.

- 01 00 00 00 is a DWORD value, which is not modified from the transmitted Length value, but could have been.

- 01 is a UCHAR (= 1 byte char) value, equal to 1, which indicates the device supports VERIFY_PIN. If Length had been longer, for example X, we would have read X UCHARs here.

IFDHControl ([in] DWORD Lun, [in] PUCHAR TxBuffer, [in] DWORD TxLength, [out] PUCHAR RxBuffer, [in,out] PDWORD RxLength)

This function performs a data exchange with the reader (not the card) specified by the slot number, Lun, abstracting functionality such as PIN pads, biometrics, LCD panels, etc. The transmit buffer length TxLength, the buffer itself TxBuffer, and the maximum expected response buffer length RxLength are also transmitted. The response buffer RxBuffer and the actual response buffer length RxLength are returned.

Transmission: 08 00 00 00 00 1B 00 00 00 00 20 00 00 08 30 30 30 30 00 00 00 00 00 82 04 00 04 04 02 00 04 09 00 00 00 00 08 01 00 00

- 08 is the CONTROL function code (defined earlier)

- 00 00 00 00 is the DWORD value of Lun = 0

45

- 1B 00 00 00 is the DWORD value of TxLength = 27 (1Bh)

- 00 20...00 is the UCHAR[27] value of TxBuffer

- 08 01 00 00is the DWORD value of RxLength = 264 (0108h)

TxBuffer is defined, by Ludovic Rousseau's proposal to send PIN remote requests (i.e. "VERIFY" APDU then CCID PIN verification data structure).

Reception: 08 00 00 00 00 01 00 00 00 00

- 08 is the CONTROL function code

- 00 00 00 00 is the DWORD value of Lun = 0

- 01 00 00 00 is the DWORD value of RxLength = 1

- 00 is the UCHAR[1] value of RxBuffer = 0, which means PIN entry was cancelled

Note: it is not yet specified how the return buffer RxBuffer should tell the upper layers what the remote PIN entry result is. Therefore a simple protocol was used: if RxLength = 0, the PIN was correct, if RxLength = 1 and RxBuffer = {0x00}, the PIN was cancelled; otherwise, the 2 bytes R-APDU returned by the smart card are simply forwarded in RxBuffer (in the case of our applet, 0x63Ci with i being the number of PIN tries left).

Appendix E – Bluetooth Pairing

The follow material gives an overview of Bluetooth Pairing. Bluetooth pairing is basically a process that consists of exchanging passkeys and setting up a trusted connection between the PDA and the smart card token. Though currently not implemented, since the Bluetooth protocol stack used (i.e., BlueZ) does not support sufficient development interfaces, the discussion gives an overview of how this could eventually work.

The PDA searches for Bluetooth enabled tokens in the area. The tokens must be set up to be discoverable when other Bluetooth devices search. During the discovery process, discoverable devices usually broadcast what they are (such as a beacon, a printer, a mobile phone, a handheld, etc.), and their Bluetooth Device Name. Depending on the device, its Device Name may be able to be changed to something more specific. If multiple Bluetooth devices are in range, and they are all discoverable, identification helps to select a specific token from other devices.

Once the token is enrolled it can toggle off the discoverability setting, since the PDA retains the address of the token. When discoverability is off, the token does not respond when other devices search for it. Undiscoverable devices can still communicate with other Bluetooth devices, but they must initiate the communications themselves, if not paired with the device.

After selecting the token, the PDA prompts for a passkey or PIN, which is shared by both devices to prove that their respective owners agree to be part of the trusted pair. With more advanced devices, such as mobile phones, both participants must agree on a passkey and enter it on each device. With other types of devices, such as hands-free headsets, where no interface exists for changing the passkey on the device, the passkey is fixed. For such devices, their associated documentation provides the default passkey, and how to change it, if possible. Often, the passkey is simply zero.

Once the passkey is entered on the PDA, it is sent over to the token for comparison. If the token is an advanced device that needs the user to enter the same passkey, it asks for the passkey; otherwise, the token uses its standard, fixed passkey. If the token's passkey is the same as that entered by the PDA, a trusted pair is formed. Each device automatically accepts communication from the other, bypassing the discovery and authentication process that normally happens during Bluetooth interactions.

www.ingramcontent.com/pod-product-compliance
Lightning Source LLC
Chambersburg PA
CBHW081910170526
45167CB00007B/3219